**Bibliographic information published by the German National Library:**

The German National Library lists this publication in the National Bibliography; detailed bibliographic data are available on the Internet at http://dnb.dnb.de .

**Imprint:**

Copyright © 2014 GRIN Verlag, Open Publishing GmbH
Print and binding: Books on Demand GmbH, Norderstedt Germany
ISBN: 978-3-668-09152-8

**This book at GRIN:**

http://www.grin.com/en/e-book/310125/contingency-planning-hurricane-sandy-incident-command-review

**Dakota Duncan**

# Contingency Planning Hurricane Sandy Incident Command Review

GRIN Publishing

**GRIN - Your knowledge has value**

Since its foundation in 1998, GRIN has specialized in publishing academic texts by students, college teachers and other academics as e-book and printed book. The website www.grin.com is an ideal platform for presenting term papers, final papers, scientific essays, dissertations and specialist books.

**Visit us on the internet:**

http://www.grin.com/

http://www.facebook.com/grincom

http://www.twitter.com/grin_com

Hurricane Sandy was without a doubt a storm that has proven the strength, dedication, and effectiveness of the emergency response operation system. Emergency responders, emergency managers, public service agencies, power and light agencies, and the US National Guard all executed their response capabilities swiftly and with great accuracy. As with most large scale disaster or state-of-emergency operations there tends to be a great amount of chaos and confusion within our response capabilities, however Hurricane Sandy was well prepared for. Hurricane Sandy is by far one of the largest and most powerful hurricanes to have ever hit land. According to CNN timeline library "the total death toll total of nearly for 117 deaths in the U.S. and 69 more in Canada and the Caribbean". With this massive Hurricane moving towards the US preparations and monitoring of this storm became paramount to all levels of government and the public.

On 'Monday, October 22 NOAA National Weather Service issued public advisories throughout the day to inform that Tropical Depression 18 had officially become Tropical Storm Sandy with maximum sustained winds of near 40 miles per hour" (FEMA, 2012). With the warnings released to the pubic of the expected effected areas the preparations and utilization of resources are underway. FEMA had acted accordingly with warning the President of the United States, mobilizing emergency operation commands, mobilizing liaison officers to the areas expecting the Hurricane, and sending viable resources to staging areas. Medical teams, power and light teams, rescue operation teams, the National Guard and the Coast Guard were all on stand-by and awaiting the signal to move in. Unlike Hurricane Katrina, Sandy's response efforts were fast, well prepared, and well equipped.

Warnings from NOAA and the National Weather service remained in effect for areas expected to be hit by the Hurricane. With social media and new technologies being used daily the "New Yorkers were encouraged to download the free Red Cross Hurricane App for mobile devices to have real time hurricane safety information at their fingertips. The app can be used to receive weather alerts and get information on Red Cross shelters. The app also features a toolkit with a flashlight, strobe light and alarm, and the one-touch "I'm Safe" button lets individuals use social media outlets to tell family and friends they are well" (FEMA, 2012). This usage of social media was one of the first times that social media has been used and proven effective by the public service agencies in warning and locating information to residents. "Along with social media usage there were an average of 10 Hurricane Sandy related photos uploaded per second to Instagram" (Webley, 2012). "Another 20,000,000 tweets were sent about Hurricane Sandy between Oct. 27 and Nov. 1." (Webley, 2012).

To expand the monitoring capabilities on "Friday, October 26 FEMA National Watch Center in Washington, D.C. elevated to an enhanced watch to proactively support any potential needs or requests from potentially-affected states" (FEMA, 2012). "October, 26 President Obama was briefed by FEMA Administrator Craig Fugate, National Hurricane Center Director Dr. Rick Knabb, and Homeland Security Advisor John Brennan on updates to Hurricane Sandy and federal actions to prepare for the storm as it continued to move toward the United States mainland. The President directed Administrator Fugate to ensure that all available federal resources were being brought to bear to support state and local responders in potentially affected areas along the eastern seaboard as they prepared for the severe weather" (FEMA, 2012).

Along with FEMA the "U.S. Department of Defense Northern Command deployed Regional Defense Coordinating Officers, and portions of the Defense Coordinating Element, in advance of the storm, to validate, plan and coordinate support of FEMA's response operations and to facilitate DOD support of life-saving and response operations" (FEMA, 2012).

With advanced warnings given, ample amounts of time to prepare and evacuated. Some residents did just that and were successfully able to survive the Hurricane due to their prompt evacuation of their homes. Others attempted to ride out the storm, although most were successful, some did not survive the Hurricanes deadly force. As with any mass evacuation the highways, back roads, airlines, trains, etc. were all delayed and crowded.

In "Washington, D.C. on October, 27 FEMA's National Response Coordination Center is activated in preparation for Hurricane Sandy's landfall" (FEMA, 2012). Throughout the Hurricane agencies such as the Center for Disease Control, U.S. Health and Human Services, U.S. Agriculture Department, Food and Drug Administration, and the American Red Cross all released information, tips, safety advice, and official warnings to prepare area impacted residents who choose not to evacuate their homes. "The American Red Cross chapters mobilized hundreds of disaster workers, readied shelters and continued to coordinate response efforts with community partners. Red Cross workers also finalized preparations of relief supplies such as; cots, blankets, ready to eat meals and snacks and were moved into place to support sheltering efforts" (FEMA, 2012).

With Hurricane Sandy being the second most costly storm in U.S. history what is the damage? The amount of lives lost due to the storm far outweigh any amount of monetary loss. CNN released "the total death toll total of nearly for 117 deaths in the U.S. and 69 more in

2

Canada and the Caribbean" (CNN, 2012). Financially speaking "Sandy will end up causing about $20 billion in property damage and $10 billion to $30 billion more in lost business, making it one of the costliest natural disasters on record in the United States, according to IHS Global Insight, a forecasting firm. The New York City mayor's office in late November estimated total losses to the city to be $19 billion and asked the federal government for $9.8 billion in aid for costs not covered by insurance or FEMA (Sharp, 2012). The theater mecca had cancelled 49 Broadway performances costing over $8 million in ticket sales alone" (Webley, 2012).

The response efforts to Hurricane Sandy was able to be promptly executed due to the advanced warnings and preparation that was made ahead of the storm reaching landfall. Nearly "8,100,00 homes that had lost power with the outages affecting 17 states, as far west as Michigan" (Webley, 2012). Massive losses such as these across multiple state jurisdictions ended up bringing over "57,000 utility workers from more than 30 states and Canada to New York alone to assist with bringing power back to the city alone" (Webley, 2012). In preparation for the response efforts "the National Guard had posted more than 1,900 personnel on duty across all the states within the storms projected path" (FEMA, 2012). Alongside with the National Guard the U.S. Army Corps of Engineering "mobilized temporary emergency power resources to stage at the Incident Support Bases" (FEMA, 2012).

Disaster response efforts without a doubt are going to require Emergency Medical Services at full staff due to the large amount of critically injured and even minor injuries people will sustain. Medical support is paramount during a disaster response operation and City of New York understood the need to be able to provide medical attention to their citizens and "had requested over 139 ambulances to be positioned within the city and an additional 211 ambulances deployed fallowing the storm" (FEMA, 2012). New York did not stop with ambulances, "Two 50-people Disaster Medical Assistance Teams were also deployed to provide basic care" (FEMA, 2012). Hurricane Sandy was definitely well prepared for by the states in its path. As with any major storm there were some areas that could use improving, however each state in the storm's path was ready, well supplied, and responded to the destruction with full force.

Hurricane Sandy hit the Caribbean islands first and then moved upward towards the Florida Keys and Jupiter, Florida. "Sandy began as a tropical wave in the Caribbean on Oct. 19. It quickly developed, becoming a tropical depression and then a tropical storm in just six hours. Tropical Storm Sandy was the 18th named storm of the 2012 Atlantic hurricane season.

3

It was upgraded to a hurricane on Oct. 24 when its maximum sustained winds reached 74 mph (Sharp, 2012). Hurricane Sandy hit the Caribbean island Jamaica on October 24 first, once the storm returned over open seas it increased to a Category two Hurricane hitting Cuba on October 25. Hurricane Sandy occasionally varied between a Category one, Category two, and a tropical storm throughout its duration. "Hurricane Sandy made landfall in the United States about 8 p.m. EDT Oct. 29, striking near Atlantic City, N.J., with winds of 80 mph (Sharp, 2012).

With Hurricane Sandy stretching across the East coast, the Caribbean, and Canada its easy to determine that there were multiple types of rescue operations, multiple types of tools and resources used, and multiple types of non-government agencies used in the rescue of victims. With the NIMS and Emergency Response Framework only have been in effect since the 9/11 attacks the system was still fresh in everyone's training and management framework, The implementation of the National Response Framework was well executed from the Federal and State level management agencies and proved to be very effective in its operation. As there are always local, tribal, and neighbors-helping-neighbors some of the rescue operations and relief efforts are not easily found or not documented entirely. Citizens now days are usually so busy and live complicated they never make time to get to know their neighbors, however when a large disaster occurs communities pull together and help each other, look out for one another, and help complete strangers in the midst of their own time of need. Search and Rescue efforts involved everyone from federal response teams, local fire and EMS personal, private EMS services, local volunteers, Military and National Guard members, and many others.

The American Red Cross mobile phone application helped with locating and rescuing victims, the app would allow for trapped survivors to be able to communicate their SOS and locations with local rescue commands. The app also kept individuals updated on ongoing hazards, operations in progress, and other important information. Hurricane Sandy was a great opportunity for not only the American Red Cross, but all public agencies to actually see how technology and social media can assist with communicating to the general public, aid in rescue operations, and keep a transparent image for the community. The utilization of technology and social media were greatly accredited to the successful rescue operations of Hurricane Sandy.

With there not being any exact number or statistic of how many successful rescue operations or how many lives were saved during Hurricane Sandy it is impossible to say how much good there was compared to all the negative outcomes of the Hurricane. As with many public safety statistics our failures are typically known and reported, rather than our successes and accomplishments. Hurricane Sandy is no different, there are many statistics on the loss of life's and property and little on the successes. The forward planning and preparations for Hurricane Sandy definitely played a large role in the response and recovery efforts and without a doubt saved many lives and assisted in the recovery tremendously.

As with everything we do in disaster recovery and incident management there are changes that need to occur as a result of the storm and emergency response efforts. Every agency that responded to the disaster rescue and relief operations will have their own inner flaws and changes to make within their organizations. On a larger scale there are a few flaws that could be corrected or readdressed. According to Dr. Redlener there were a few main issues that should be addressed In regards to Hurricane Sandy, the first being evacuation procedures and post-incident communications. "More information and planning are needed to effectively prepare for and implement general evacuations as well as critical, complex evacuations from hospitals and health care facilities" (Redlener, 2013). The notification to the public to evacuate was acceptable and useful, however with only a few roads and highways leading out of the suspected disaster area there were a multitude of accidents, traffic jams, and delays in the evacuation process.

The hurricane Sandy incident command systems ultimately did a good job at adapting into the National Response Framework and establishing communications with local, state, and federal agencies, however the post-incident communication fell short of adequate. According to Dr. Redlener, "We need better after-action analysis and more research in order to ensure sound strategy, policy and decision-making" (Redlener, 2013). Dr. Redlener also noted a few other topics in his forum "A Critical Catalyst: Lessons Learned from Hurricane Sandy" such as; Mental Health concerns, Citizen Readiness, Training funding, and Vulnerable population evacuation.

The lessons learned from hurricane Sandy are quite obvious and fairly easy to address. The overall incident command and National Response Framework applications during the hurricane was executed very well. The issues in regards to evacuation of vulnerable populations and mental health issues that could arise later down the road are going to have to be addressed at the local level and will take time and collaboration to find better suited solutions for that specific area and issue. In regards to the post-incident communication issue there are many ways that this can be corrected or better suited to meet all stakeholders request.

There are a few very important key elements that we can all learn from hurricane Sandy. The fallowing are, but not limited to, some key points to summarize our findings from this incident;

1. Social Media and Technology are very valuable and effective tools for communicating to the general public and conveying accurate, real-time information.
2. Communication between all agencies fallowing the incident is as equally important as communication during the response and rescue efforts of the incident.
3. We in public safety need to place a greater emphasis on educating and preparing citizens to be prepared in taking their own initiative to plan, educate, and prepare for disasters.
4. Establishing plans and agreements with local businesses and organizations to work together during times of a disaster and collaborate ideas, training, and resources.

Although these are not everything that can be learned from Hurricane Sandy, they are key points that have been identified as needing improvement. As with every large scale disaster there will always be issues, chaos, and needed improvement. These real world applications allow us to further identify and improve on our existing policies, action plans, and incident command systems.

Works Cited

*A Critical Catalyst: Lessons Learned from Hurricane Sandy* Published January 30, 2013 Hurricane Sandy by Dr. Irwin Redlener

By CNN Library Updated 2:37 PM EDT, Sat July 13, 2013 http://www.cnn.com/2013/07/13/world/americas/hurricane-sandy-fast-facts

*FEMA* website http://www.fema.gov/hurricane-sandy-timeline

Livescience.com Tim Sharp, Live Science Reference Editor | November 27, 2012 10:50am ET

*National Hurricane Sandy by the Numbers: A Super storm's Statistics*, One Month Later By Kayla Webley @kaylawebley Nov. 26, 2012 Read more: http://nation.time.com/2012/11/26/hurricane-sandy-one-month-later/#ixzz2ebxMGU9q